Mariana

DEDICATED TO …..My loving and supportive mother, family and friends.

Description

My novel is a family oriented SCI-FI that is based on a family named the "Mays" who have insight on how to provide human life with hope for future existence rather than extinction. Save our human race…….."SOHR".

Bio

My name is Garwick Stevens; I was born in Montreal, Quebec Canada 1968

I am a brand new author with wild ideas. I was always told as a child to write things down because I had the talent to create crazy stories.

My idea for my book came to me when I was hospitalized for mental illness in 2009.

I hope you enjoy reading my first novel, "MARSANIA"….."A CHANCE TO SOHR"

Thanks for taking an interest in my first Sci-Fi novel.

Warmest regards,

GARWICK STEVENS……..

"THE TABLE OF CONTENTS"

Chapter 1

SIMON MAY IN: ……"MARSANIA"……"A CHANCE TO SOHR"

The year was 2,000,000 BC when earth was inhabitable.

I am Simon May, and I am a young prince (in a small sort of way) with a big idea for human life. My dad Graham May is King of the people (unofficially). He plans to be the ruler of Mars someday; a planet currently ruled by the people. In the universe, every race has its planet. The people look much smaller and smarter than people today. There is an air about them, the way they carry themselves.

My mother was sadly killed on a bounce to Earth at the young age of 23. An unfortunate incident happened where a dinosaur trampled her. Her death came shortly after my brother Stephen, and I was born. Her name was Val Snow. She was never married to my dad because they had irreconcilably different beliefs in life.

My brother and I bounce to Earth sometimes, in a porthole with boots that would travel from planet to planet faster than the speed of sound. Here we would hunt dinosaurs that are incredibly huge and are very scaly in shades of grey. We would eat the meat and share it amongst all the people on mars because nothing is growing on mars. The planet sadly is dying and has barely any life because we are killing it off slowly. We also find entertainment in the hunting itself. While on Earth we need regulators to breathe the air because it is too warm and heavy. Only the dinosaurs and plant life can breathe on earth.

I have a wife, and her name is Criz May. We've been together for 8 enjoyable and painful years. She is the mother of Mickey May, our son. He would have been 8 years old today, September 2nd. Mick always loved it when we would take him on a bounce and teach him how to hunt dinosaurs. One time I promised Mick that I would take him to see the Egyptian pyramids that have been there as long as I can remember. They are a gathering place to party for all the people within the universe. I took him when he was about 6 years old. He didn't stop talking about it for the next two years. The curiosity and fascination on his little face would have brought a grown man to tears.

When we went on a hunt we would not just kill the dinosaurs with one shot, we would track them down and taunt them with weapons of mass destruction. This ritual was extremely entertaining for everyone involved. In my family, we all love to hunt this way; it is 'the way of May's. My best hunt ever took over 10 days. That dinosaur was a marathon runner. It just took off and was too fast and difficult to follow, but we found it. To Cruz's excitement, I let her make the final kill on that one. This is a ritual that we carry out after the beast is dead: we take it to Stonehenge another party place that has been there longer than the pyramids, to prepare it for the feast, We sometimes swim in the insides of the beast just to feel a thrill. One day my father and my son Mick went on a hunting trip. I remember this day clearly, and it is one day that I will never forget. My son Mick can't differentiate between his left and right foot, so frequently his mom Criz would help him with his shoes. On this day he refused to have his Grandpa help him, and he ended up putting on two left boots. Unaware, Grandpa watched him bounce up and get ripped in two while on a bounce to earth. They were supposed to attend to a kid's birthday party, Jack Moore, he was turning eight.

Cruz can't get over the guilt of not being there to help Mick. This has torn apart my insides and destroyed my life for more than a year. I haven't bounced since, but I know that I need to do one day. We need to inhabit the earth. The scientists will need my advice on where to put the weapon to change the air and life, to enable us to breathe and live on earth. We need to do this soon so our people can live on Earth forever; we need to change the atmosphere fast. This will take a year or two to complete. We have already been working on this for several years. The name of this bomb will be SOHR; standing for 'Save Our Human Race.' We will call it just the Bomb.

Chapter 2

DINNER

It was getting near suppertime, and we had a lot of dinosaur meat to eat like brontosaurus and t-rex along with some fruit. The hunts feed all of the people on Mars. My dad had just walked in, and everyone stood up and said, "We are the people of Mars." He finally sat down, and we were ready for the feast as we started to argue back and forth and repeating ourselves. My dad speaks for all of the people on Mars, and he turned to me and said, "We need to move fast. The weather is changing on Mars, and all forms of life are dying my son." I said, "I know dad. The people are working on it."

My wife whispered to me, "The people are getting nervous, my love. We have to do this soon." Then she cried out with more urgency "Simon my love, we need this done soon, real soon!"

I replied, "All we can do is wait until we get word from John that the bomb is ready. My dear family, we need to do this soon, I know. I can feel the weather changing for the worst." John is the scientist responsible for building the bomb SOHR.

My Dad agreed, "Yes! My people, we need to do this soon, or we will all die on Mars."

We finished supper at last, and after a long talk, we knew we needed to work on getting things done quickly. The people were getting extremely edgy, getting ready for the entire population to bounce. I don't think that all of us will clear the planet in time, seeing as only one person can bounce at a time. But I can't focus on that right now.

I said, "Dad, I am going down to talk to the scientists." I would finally get to see how the bomb was coming along. I took a bounce to get there. It took less than a second; it's way faster than walking.

 John is the man I only talk to about the plans for SOHR, so that they are precisely precise, right down to the last detail. John said, "Hey Simon how is the family?" I replied, "Scared John! The people are getting nervous. I need to know how much longer we must wait before we can set the bomb."

John said, "We are looking at a week or two more, but we are not completely positive that this thing will work. Simon, all we can do is pray hard, and try our best using all of the information we have. We will do this for Jim, so his legacy will live on."

I was getting more anxious about the completion of this project, "What are the chances of getting this done today? Can we place the bomb today?"

John replied, "Yes, we can, but I am not sure if it will work. Or if it will kill all of human life for good! Do you remember the other planets, Simon?"

Jim my distant relative, tried to inhabit planet Earth and failed miserably, he was the founder of SHOR. I decided to bounce back to my father at the tower. "Dad, we can do this sooner. I have a good feeling that this is going to work! "

"Son this *is* going to work. We don't have a choice. It has to work. My people are looking at our last bounce as their last hope." He looked out the window, at the near-dead planet Mars with a tear in his eye. "We won't stop until we succeed. All of the others are depending on us."

Jim! Help us so that our efforts don't go unwarranted. Make SOHR successful. I silently prayed that my father will be remembered and loved for eternity and *prayed for God for his blessings.*

Chapter 3

MY WIFE'S LOVE

Morning came, and I was getting dressed to visit my dad. It felt like today would be my first hunt in years since my son died. My dad was downstairs already having breakfast with my brother, My wife, and I are usually there first, then my brother, and then my father. But today my wife was still sleeping.

I was going to bounce to earth for a hunt, and I was excited. For the first time in a long time, I felt a rush come over me. I felt as if my brother Steven and I would be lucky enough to get something big on this bounce. I sat down and asked, "How are things, Dad?"

My dad replied, "Ok, how are you, Simon?"

"Ok. Stephen and I are going on a hunt today, and to take a look around at our future home."

My brother added, "How do you feel brother? It's been a long time. Are you feeling good about this bounce? I'll help you get there, we can do this together!"

"Ok," I said with a smile "OK, but I need to talk to my wife before I do this bounce. I'm thinking of seeing my son's grave with Cruz."

I was ready for a feast. We all eat well on Mars; no one is poor. Everyone gets food from all of the hunts; it's a sound system for all, and it works.

After breakfast I went through the long dark and wet hallway; without a bounce, this takes too much time,

No one walks on Mars; we all bounce From place to place, it's faster.

I was looking for my wife. I figured she was still in the bedroom. I reached the bedroom door and gave a small knock before I opened it. I walked in slow and quiet; she was still sleeping peacefully. I lay down beside her and kissed her as I whispered "Hello. You're late for breakfast. Are you feeling okay, Criz? Are you hungry? Can I bring you something to eat baby?"

Cruz shook her head, "No food, I am just thinking of our son, and your mom, how they died. Simon is this too much to handle said: "No, do as you please my love, I love you."

After lying there all morning in bed, we both were in tears thinking of Mickey and Val, and what happened that day. We always think what if? If it didn't happen, what if he was still alive! Criz cried out loud "My baby May, please come home". And then Criz cried hard. This was the worst I had ever seen Cruz. I was hoping she'd get over what happened to my son and mother, or she might have to see our doctor again for some medication to get over the death of our son and some pot to relax. This pot-smoking works for forgetting things that hurt. The pot is a brain eraser, and it relaxes the body. It works and it's illegal on Mars but we can smoke it on earth too. Everyone smokes pot, even me. It helps to deal with deep issues like death. It stops the crime a little bit; when people smoke they don't seem to care about committing crimes.

After a long day, we finally got to my son's grave, both in mourning and with tears and love on the wind. We were both crying, our eyes are full of tears, as I hugged

Criz and said, "We love you little May, we miss you, little guy."

Chapter 4

THE BOUNCE

It was getting near supper again and my dad and I were at the lab talking to John about the bomb, and where to place it. I said to John and my dad, "When I am on my hunt, I'll try to see how and when to place it." My brother just bounced into the lab out of the blue and was breathing heavily. He looked like he was messed up on drugs; he loves the feeling of being high when he speeds through the bounces. I yelled, "Steven! What the hell? Can't you knock?"

He handed me a joint and said, "What the hell? Just smoke it."

"John. How much time do we have before the end of mars?" I asked.

"We do have a few years my son, but we need to start this project now. The life of all humans is now hanging on our shoulders."

"John!" I said, "We are going to find a place on earth for this bomb. It will change the atmosphere, right? So we can populate the earth."

"YEA!" My dad interrupted. "We need the earth," he said. "This planet is dying fast," he added as he looked out the window at a red sky. He could see another rainstorm coming through. "My son," said my dad "Please make sure you find a safe place to put the bomb, in the right part of the earth, like the North Pole!" He then jumped out on a bounce and went back home.

Dad was now in the dining room. My brother was on the other side of the room, looking at all of the freaky animals that the scientists had brought back from earth. They were laughing and poking the animals with a stick.

I yelled out "What the hell are you two doing with those fucking things? They are going to die soon! They have no air dummies!" I started to laugh like I haven't laughed in years. Gentlemen!

These animals caged before we were the only species that could breathe on Mars; however, after a short time, they would die, slowly, like fish out of water. They look similar to a fish of sorts, but also have legs, and arms with claws that can cut like razor blades. Their body mass consists of only fat, making their only defense - razor-sharp claws. We have to stay away for our safety because they will kill in a matter of seconds, usually going for the jugular vein if threatened. For this particular reason, we call them 'Jug bugs'. If you see one on earth you must come to a dead stop, fall to the ground, and grab your neck so it will only cut your hand, not your jugular. This is the only planet left we need it, to carry on human life and life of these hilariously ugly creatures.

Chapter 5

EARTH: THE HUNT

While on my bounce to earth, I felt like wearing two left boots, but I couldn't do it 'because of my wife Cruz. I think she might have thought I would like to do this. But Just the thought of dying and leaving everyone stopped me from doing such a dumb thing. So, the thought was there, but my brother Stephen made sure I was getting the right and left shoes on correct. My son died on a bounce because he wore two left shoes.

I finally got to earth and started to look around for a good place to set the bomb. It was cool in some places and hot in others so we needed to wear ventilating masks. If we didn't wear the masks the air would burn our throats; it is too hot when we breathe it in. We decided to stop on the hottest part of the earth. They call this place Africa. It looked like a good place for John to set down the device, and prepare for the blast that would change the atmosphere.

It's different here on earth, the air is different, and the plants are huge and bright green.

When we got through the bounce there was a dinosaur right there; we decided to leave it, and not kill it 'because it would make the hunt too fast and not fun. I dropped a fuzz bomb, which made a loud 'fuzz' sound like shoeing a cat off, which scared the thing off. It ran so fast, like a cat jumping out of a tub filled with water. This was the funniest thing to see. We use these bounce shoes to fly through the air to find a wounded dinosaur, which is easier to kill. We use stun guns, which give it a small heart attack, and then we follow it down for the kill. Sometimes this takes a few days, but sometimes we find a dinosaur that is much wounded that dies fast. We look for the long kills for the fun of the sport.

My brother was moving fast through the bush with a dinosaur in mind; I think he can smell it in the bush.

I asked him, "Stephen what do you see? Or shall I say what do you smell?"

Stephen said, "I smell a large one, but I think he is too wounded to follow, let's bounce." So we bounced a half-mile in the air until we saw a large one for us to hunt. Stephen said, "This will be the one bro."

I said, "So cool!"

Stephen replied, "So cool! So cool! This is the one for us to hunt."

I yelled, "Let's do this one!" We swept down behind this beast and started the hunt. The back is the safest place to stun these things, cause if we were in front of it, it would charge at us too fast and we would not be able to get out of its way.

"Shoot now Simon," Stephen said. So I did. The stun made the dinosaur drop for a few minutes and then it took off fast. We followed it in the air, as it moved super-fast. Stephen took another shot. The beast fell after that shot and then got up and ran. This would be a fast kill. I think the beast already had a heart problem. It took 2 hours for us to hunt this beast, in the air and also on foot. The best shot was right between the eyes; it made the beast look unreal and funny. By the time we killed the beast, it was getting late and we needed to get to Stonehenge soon. This was our last kill, to feed some of our people on Mars. This hunt was helping to take my mind off of my son, as well as helping to find a new home for the life of man. I was ready to help and take on Mars, and help the people to relocate here on earth as soon as possible.

Well, we finally got our kill. Stephen said, "Let's party you animal! We both deserve it you kooky. Let's get to Stonehenge, I need to smoke a joint".

So, we took the beast on a moving platform to be dissected, and to get it ready to eat. I said to my brother "What a long ride, I need to smoke one too. It's been a long time, so come on and let's get high and smoke a joint or two". We bounced the hell out of there and talked all the way through to Stonehenge for the party that would never stop.

"STONEHENGE"

Chapter 6

THE PARTY

After the mission was complete, Stephen and I decided to celebrate. We bounced around planet Earth and landed smack dab in the middle of a huge party with a live band on stage that kicks ass. It was at a place we called Stonehenge. There happened to be a lot of people at Stonehenge. It was one of the most popular places in the universe where it is very common, almost expected of a guy, to get real high, and loud, and to get all messed up and sloppy drunk.

I walked in with my brother to the left side of me. Then right there in front of me, stood a girl I recognized from my past. She was a Victorian princess. She was not a royal princess, but she was a protector of secrets. Victoria is her name. She has a very hot body, blackish hair, and super huge boobs and is wearing a tight dress, I knew her before I was married to Criz, my wife. This girl had always loved to party, and all that I ever wanted was a family instead of the party life. I am a little bit shy to speak, but the party instinct inside me took over.

I said "Victoria? Is that you?" She did not hear me, so I went over to her and in a loud party cry I shouted, "What's up Victoria?

My brother cried out like a wolf that had lost its pack and took off to get screwed up and crazy drunk like a wild animal.

She whispered in my ear, "I love you Simon May; where the hell have you been for all of these years? I know you are married and stuff Simon, but this girl here would love to be your part-time, on the side."

Nervously, I asked, "Do you want a drink?"

"Ok", replied Victoria. "Can we go to another room and spend some time together?" she asked.

Without a second to gather my thoughts, Victoria grabbed me by the shirt and dragged me to a dark corner. My brother showed up and pulled me out of the sticky situation.

"Look who I found," said Stephen. It was a guy I had also known in my past. But I had forgotten his name. Stephen said "bro, do you remember Todd? The man of the year! This man was going to marry Criz when she was 18, and her mom insisted this was not the guy for her."

All of a sudden Todd hit the floor and passed out. We all yelled out as if barking at the moon like a pack of hyenas. The crew took Todd to the sleep room, where we all go when in that wasted state.

Laughing like some kind of crazy person I said, "I haven't been messed up like that in the longest time." I walked through a crowd of crewmembers and fallen down, passed out, wasted patrons. Making my way over to this girl that had been trailing my ass for the entire night, I asked her name. "Hi sweetie, you look good tonight."

She said, "Hi, I'm Holly, and you are the reason for me being here tonight."

"You look good too," I said, "Let's dance."

So holly and I busted a move amongst the unconscious bodies piled on the floor. We made the music dance. After a while, with the rush of the music, Holly said, "Let's get a room!"

I decided that it would be an extremely bad decision if I had gone with her, so I left her standing there alone, I was feeling extremely parched. I went back down to the bar, and I noticed that most of my childhood friends were partying in the corner. I yelled out "Hey you guys, party hearty!" They all yelled back in sync "Wow! Look who it is, Simon May!"

I got myself a drink and wandered past a few nosey people staring at me, as I looked for a place to sit. While grinning at them, I wondered what the deal was. Could it be my name? Am I a wanted man or something like that? As I walked by another room, I heard a burst of joyous laughter. I went to the room to find out where it had come from. "Hey! It's Bailey from Venus!" I shouted out "Bailey you rock!" As I went to sit down, a glass had smashed on the ground. All of a sudden I woke up in a bedroom with blood all over my body, as I remember only one thing, swimming inside of a dinosaur, for the thrill of the kill.

Chapter 7

THE DAY AFTER

It was one o'clock in the am. I had just waked up beside Bailey she is one of my brother's best friends and is a knock out chick. I'm all bloody from the night before; this is not a good thing at all. I was thinking about my wife, and I didn't know where I was. I didn't recognize any of this stuff. I thought maybe I was on Mars or Earth. I still had my clothes on, which was a good thing. I think I am on Mars or some other planet? For my sake, what have I done?

I could use a giant, ice-cold glass of water. I decided to get up and grab some water and a cigarette too. This would help a lot, to clear my head. Water always helps after a blackout drunk. All I had to do is wait till Bailey wakes up.

I didn't want to look outside, afraid of what I might see. I might be on the moon or some other planet, so far off. This scares me a lot, not know where I was or where I have been. I never know what to expect when I wake up disoriented. I whispered to myself quietly, "Where the hell am I?" The number of days I might have been on earth was unclear, what a party. This also scared me. And the blood on my clothes, "What the hell?" I said as I looked all around for a clue. "Well I am not black and blue, that's for sure", I muttered as I looked into a 'mirror of fright' on the wall. All I needed to do was wait till Bailey gets up from her sleep. I got back into bed beside her and waited for her to come too, trying to remember what went on during this blackout drunk.

Bailey started to wake up a little; this was still going take some time. She was still pissed drunk, and I think she was partying on earth for a few days before I got there. I meet Bailey through my brother Stephen at a party on Earth. This chick is hot stuff, but I haven't seen her for 15 years. We never had a sexual relationship; we didn't even date each other. Strictly friends Bailey and I were.

All of a sudden it came to me that Bailey was from Saturn. "Wow!" I said as I took a huge gulp of water and a long drag from my cigarette stick. I could see a pile of mail that she had left on her desk; it looked like she had not opened any mail in months. Slowly my mind was coming back.

I was so afraid of what I might have done the night before, never again! Just one day at a time is the way I would live starting knows, the sky was cold outside, just as I remember it on the planet Saturn. I am not that far from Mars. Just one bounce is all it will take to get home, twenty minutes or so depending on the number of other people ready to bounce. We must wait for enough air space before we can take off. It was slowly coming back to me, all of the events that occurred before I woke up in Bailey's bed.

It had been two hours since the last time Bailey moved. Thanks for the grown of Wonder! She was still breathing; I think she shot-up with some heroin just as she got back from her bounce on Earth. I didn't use any of that shit that I can recall. I knew, 'because I didn't puke up the dragon; I knew for sure that I was clean from heroin. I didn't feel the dragon at all inside my stomach or body. I said to myself "Thank you! I am clean from that stuff! I don't like that junk in my body at all! It makes me sick".

I decided to have a shower to clean up and get the stink and blood off me. It took me an hour to drop a load, shower, and shave. I needed to get out of this place and get back to Mars; I was missing my wife a lot and it was making me feel a little sick to my stomach. Just as I got out of the shower, someone pounded on the door. It startled me, but it could have been important so I needed to open it, just a crack.

It was my brother Stephen. I opened the door wider and said "Hey bro! What's up? Why are you here? How did you find me?"

Stephen said, "I know where you are at all times, Simon."

I replied, "It's a little strange bro, you keeping tabs on me.

Stephen came in and sat on one of Baileys chairs, just glaring at Bailey on her bed. "What a messed up person she is Simon. You didn't do anything with her, did you?" Simon said while he laughed hysterically.

"No way!" I said. "I just woke up and there she was. Hey dude, do you know what happened in my black-out drunk?"

Stephen said, "Yea! You were sitting in the corner of the bar with Bailey, and someone brought in a jug bug and it got loose. It started to kill a crapload of people at the party. The entire bar went crazy to get away from the jug, and you ran face-first into the wall and knocked yourself out. Face first, bro!"

I said, "Wow that was a good party. How many people did it get?"

Stephen said, "At least twenty-nine people died, and ten were injured. What a mess Simon. People were running around trying to kill the jug-bug for at least thirty-seven minutes. It was too fast, and the people were too drunk to get a good shot in, but they finally got it!"

I said, "Wow! How close did it get to me? I mean did it try to get me?" I was so happy to still be alive.

Stephen said, "Holy cow! Am I a lucky bastard too, or what?" I asked my brother Stephen to get me out of this place, and to help me figure out what to do about Bailey.

Stephen said, "Let's get out of here and just forget about her."

So I grabbed my stuff and put my boots on, then just as fast as the speed of light, we were on our way back to mars.

Chapter 8

BACK ON MARS

Well, I got back to Mars and everything looks okay, but the atmosphere seemed a bit different; the skies weren't looking any better. I got off the bounce with my brother Stephen and he left me to go and see John the scientist.

I am off to see Cruz. I got to the door and it was locked so I tap softly. An old friend of Cruz's answered "Jasmine"! She has been in our family for years, since Criz was a child. I like the girl - she is a good person. Jasmine has never been on Earth; she is a clean person, just like Criz they stay away from all the parties on Earth. I like this cause this means I can party as I wish. As I walked inside to our open-concept apartment and I said "Hey Jasmine! How is it going? Been here long?"

Jasmine said, "No just a few days, how was earth and the hunt?"

I said, "It went well. I killed what I could and then Stephen and I went to Stonehenge for a party, and I got all messed up and nearly got attacked by a jug bug."

Cruz came into the room and said, "Simon, are you okay? Did it hurt you at all?

I said, "No honey, I didn't get hurt. I was lucky sweetie."

Cruz said, "Thank God! But you had a good time?"

I replied, "Yes! I missed you a hell of a lot Cruz, and I got what John was looking for, that perfect place for that device thing. I am going to see John after I spend some time with you Criz.

Just then, Jasmine came into the room and said, "Ok love birds I'll let you guys get reunited. Bye for now, love you Criz."

Cruz said, "Bye for now Jasmine, love you too." Jasmine walked out the door and bounced. Now Chris and I were alone one more time. I was going to take a shower before Criz told me to. I said, "Cruz! I am jumping in the shower before we do anything together."

After I got out of the shower, Criz was in the kitchen and there was a box on the bed with a red ribbon around it. I did not want to ask but it had my name on it 'Simon' so I started to open it. When I looked inside there was a letter. I opened it and I started to read it.

It read, "Dear Simon, I know we lost our precious son, and I love him with all of my heart, but what I need to say in this letter is that we have another on the way. Simon, I am pregnant with another baby. We need to name it, and love it just like we did for 'Mick May' Ok? This is our new life and I think it's a boy, that's what the doctor said."

I said to Criz "Wow yeah! A boy! Are we sure you are pregnant?"

Cruz said, "Hell yeah! Simon, we have a new life inside me, and I think your dad needs to name it." As Cruz came out of the kitchen and sat on the bed, she pulled me down and gave me a big hug and a long kiss, and whispered: "We need to go for supper soon, let's lay here 'till then."

I said, "Ok." We kept on kissing and hugging even longer as I rubbed her belly and was completely overwhelmed with feelings of love and joy.

Chapter 9

THE FEAST

Cruz and I got ready to bounce to the dining room and to join my dad Graham for supper. My brother Stephen should be there too. I have lots on my mind to tell Graham my dad; I can't wait to tell him.

We just bounced to the front door. My dad was there but Stephen wasn't.

I said, "Hi Dad, we have lots to tell you. I know where we can put the bomb."

Graham said, "I need you to tell John. Simon, you must tell him."

I said, "Ok, but I have another thing to tell you."

Graham asked, "What is it, my son?"

With Cruz standing beside me, holding my hand, I said, "Dad, Criz and I are expecting a baby in our family, and it's a boy, and we need you to name it."

Graham said, "I have the perfect name for our new family member; we can call him Matty May."

Cruz and I agreed, "Yes we would love to call him Matty May."

We all agreed on the name, but we still needed to tell Stephen that we are going to have a baby more time and then I asked, "Can we eat?"

Graham said, "Dig in my son, I am a happy father and grandfather now. Please fill up we have lots of meat. We have the whole dinosaur to eat."

So we started to eat without my brother. Dinosaur tastes like a cow, like beef, with honey poured on it, and with mint sauce inside for the blood; the best way to cook it is fast, so the juice inside waters in your mouth, with a glass of wine. The meat should be good and tender. A whole dinosaur should last about half a year with a big family.

The room was filled with utter silence. It got creepy that my dad was not saying anything at the table. There must be a lot on his mind with John, and the bomb, and now Matty, our new son. I asked, "Dad are you okay?"

He replied, "Yes Simon, I am okay. I just want everything to work out for all of us here on Mars, and to build a new life on Earth.

I said, "Good I am glad you are okay, please hang in their dad. We will save the human race, and life, as we know it." We continued to eat more, and drink more. I knew my dad would be okay with all of this stuff on his mind, but this was the quietest I had ever see him. At least he knew he would be a grandfather again soon. I guess it hadn't sunk in yet. My dad sat up and said, "Simon, I wish your mother could hear the news about Matty. Wherever she is I know she would love to hear about the new grandkid in our life, and Mickey would have also loved it too."

It was bothering my dad and me. I think it was on my wife's mind too. But a new child would play tricks on our minds; it's just new news for our family, one step at a time we will get through it, we always do.

All of a sudden Graham said, "I love you all, we are all in this for the best. Let's have a quiet prayer for our diseased."

So we all held hands and went quiet for about 2 minutes. Everyone had tears in their eyes when it was all over. Too bad my brother Stephen was not there.

I said, "Dig in guys, I am really hungry."

Things started to look happier after I spoke; I know the prayer did it.

I said, "I am feeling better, enjoy the feast. I love you guys, and God bless us all."

Chapter 10

BACK WITH JOHN

I had just left my dad with Criz in the dining room so I could go visit John; to tell him that my brother and I had found a few places on earth where we could place sohr the boom.

I bounced, and it only took a few minutes to get to John's front door. I just walked in. He was playing with a jug bug, poking it with a stick and mumbling to himself. It was really funny to see him doing that. These bugs were dangerous and would kill you in a heartbeat; they attack and they go for the jugular vein to kill and then feed. That's why I stood back in fright while he kept poking it, which made it even funnier.

Just then my brother walked in and said out loud "What's up?" This bothered the jug bug; it ran into the corner of the cage and John laughed.

We all started to talk about the bomb, and then John said, "We can place the bomb in a few weeks, I just need some time to test it, to see if it will all work as planned."

My brother started to poke the bug, and said, "I love these things, guys."

I backed off because the door of the cage was open. John started to laugh like a crazy man, and after listening to the laugh I let out a crazy laugh too.

I was looking around John's office, and I notice that the machines that would change the face of the earth were in the corner of the room. I asked John if they were the ones and he said, "Yes."

I took a closer look at them; they looked really big and mechanical, with pipes and gadgets all over inside and out. I asked, "John, what are the chances that these things will work."

John said, "They will work my son, I promise."

I took a deep breath and said, "Thank God." As I stepped back and praised John's work with excitement and pleasure, my mind went back to my son and I having fun on Earth hunting for dinosaurs and critters. I remembered when Mickey fell on a branch and his gun went off and shot a jug bug dead. That was a great memory. I took a bite from a sandwich that was left on the table and wiped my mouth as I looked at the machine in the corner that John had made. I took a drink to wash down the stale sandwich.

My brother asked, "What's for supper?" with a little crazy laugh.

I was standing in the corner when Stephen came over and told me that he was seeing this girl, but the only way he could visit her was with me. I asked, "What girl are you seeing?" I have big we are having a baby. And Stephen said, "What the hell this is the coolest thing I have heard all week.

And then Stephen replied, "Bailey, and she is on Saturn in a rehab center."

I said, "What? That is crazy! Of course, I will go. I love to see someone down and out." I laughed a crazy laugh and blew away John as he laughed too from across the room. The noise made the jug bug squeal.

Then John said, "I guess you guys are in for a vacation." Stephen and I both said, "Yes! Let's go on a crazy trip, do you want to go with us, John?"

John said, "Not a chance in hell! I think I will stay here." Stephen and I went out the front door and bounced to the platform that would take us to Saturn.

Stephen said, "We are on our way! Let's go crazy!"

Chapter 11

THE REHAB

It took us twenty minutes to get to Venus, the planet of rehab, on our bounce. There were two hundred people in front waiting to see their beloveds; it was a laugh to see how many people care for family and friends. We were all allowed to smoke pot on Earth but not Mars. I would only do it after a long hunt.

We had just reached the front door when Stephen said, "This is going to be crazy."

We rang the doorbell and this big colored guy let us in. He looked like he was from the moon and had just landed a job. He still has his tribal tattoos on his face. The man said, "Hi, my name is Keller, can I help you guys? And please leave your jump boots at the desk."

Stephen said, "Yes killer Keller, you can help us big guy. We are here to see a patient named Bailey, but we're not sure what her last name is, does that matter?"

Keller said, "No, I think she is on the third floor in the freedom wing. Can you guys find it okay?"

I answered, "Yes we'll just follow the signs." Stephen and I started to walk towards the elevator, and we both looked at the sign that said 'Freedom Wing' third floor. We looked at each other and laughed. Then I said to Stephen, "So we are in, are you still okay with this?"

Stephen replied, "Hell yeah! Let's do this together." We stepped onto the elevator floor that was made of glass, so we could see right through it. As we reached the third floor and the door opened, this crazy lady was screaming, "No meds."

Stephen spotted Bailey and walked away from me while letting out a party cry directed at Bailey.

I said, "Bailey is that you?" as I started to walk over to her.

She looked all medicated up and slow. Stephen said to her "Let's get you out of here; your thirty days are up."

Bailey replied slowly while drooling from her mouth, "Please help me, I need a fix". Now I know why my brother brought me here; it's easy to get some booty from Bailey.

A loud bell rang out, and everyone on the floor started to move to another room. Bailey said, "It's lunchtime."

Then Stephen said, "free lunch!" We walked towards that room acting as if we were patients and sat down beside Bailey. A nurse put a dinner plate in front of Bailey and Stephen, and I had to wait. When I finally got my plate, there was an egg sandwich on it.

I muttered, "Crap! Just what the doctor ordered."

Bailey laughed out loud. It was good to see this emotion from her. As she ate her sandwich with her mouth open, I laughed my head off.

After a short time, we finished lunch, and my brother Stephen went to the bathroom and was there for a long time. I think he was using drugs, what a jerk! Well, at least he was keeping it away from Bailey.

Bailey asked me, "Why are you guys here?"

I replied, "'Cause we care Bailey."

She said, "I love you guys, Simon and Stephen - you guys rock," as she hugged me. Stephen finally came out of the bathroom, and his eyes were red, what a jerk. It took him twenty minutes to finish up his nasty habit. I think Bailey knew what he had done in there but she didn't ask. I think she just bit her tongue and turned her head from all of it.

I whispered in Stephen's ear, "I think *you* need this place." He laughed as he went into another room with a TV in it, and more crazy people. I stood in the hallway with the doctors and Bailey, as my brother bothered the patients, making the doctors mad.

I asked Bailey what she wanted to do and she said, "I want to stay here in rehab for a while ok?"

I said, "Ok, if that's what you wish."

I then went over to my brother and told him "Let's get out of this place."

He asked, "Can we take Bailey with us?"

I answered, "No, she won't come with us. She wants this program for real Stephen, she won't leave."

So my brother and I left. We went down to the main lobby, and we got our boots back from Keller and we bounced to the nearest transport station. This bounce back to Mars would take some time because there were a lot of people waiting to jump. I figured it would take one hour or so. We finally got our turn and left Venus for Mars. It did take a long time, but we got through.

I was feeling bad about leaving Bailey as I say to myself, "Good luck Bailey."

Then Stephen asked me, "I like Bailey bro, what shall I do?"

I left him alone with that question and just laughed, "Ha," as we entered Mars.

Chapter 12

AT HOME WITH MY LOVE

We finally reached Mars and my brother was at my side. I said nicely to him, "Stephen I'm off to see my wife, I haven't seen her for a while. Please don't take this the wrong way."

He replied, "No way bro! I am off to the moon for a fight."

I laughed and walked away.

I was finally alone, as I jumped to my front door and walked in. Cruz was on the phone, and she turned and saw me standing in the doorway. She was talking to her old friend Donna. She said, "I have to go now, my love just got home from Venus, bye for now."

Before I could say hi, her arms were around my neck. Cruz kissed me on the mouth and said, "I missed you, Simon!"

I said to her, "Love you too sweetie."

We both sat down and her arms were still around me. Cruz asked, "How was your day love?"

I answered, "Stephen and I went to visit my old friend Bailey in rehab, on Venus. I was there for a long time, and I am glad that I am back with you, it's getting late and I am tired."

We both had something to drink and we went toward the bedroom to lie down. We hugged and kissed each other more. I asked Chris, "Will things work out for the people?"

She said, "Yes my love, things always do, that's the freaky part of life Simon."

This eased my mind a little. I then turned on the television to watch the news. This man was talking about Mars and the other planets, and how Earth is a big asset for new life, as we have been told. Cruz was still lying beside me and was just about to fall asleep when someone knocked on the door.

I got up slowly and went to see who it was. It was Stephen. I opened it slowly and he walked in, he had a black eye. I said, "What the hell? What happened on the moon?"

He said, "Just a fight my brother," as he went into the kitchen and opened the fridge to get a beer. He then asked me, "Where is Cruz?"

I said, "Asleep, it's been a long day my brother. What's up?"

"Nothing" he replied, as he sat down on the couch and put his feet up on the table. "Simon, I am worried about the planets."

I said, "Me too bro" as I got a beer from the fridge. I asked, "What's going to happen to our planet?"

Stephen said, "Bro I hope the thing that John is planning, works. From all of the people working on this, it looks good. This is what John said.

While I finished my beer, Stephen reminded me of the new baby we are having, and how I missed Mickey and my mom Val Snow. I hate it when he does this. So I said to Stephen, "The past is done, and now we need a future for all of the new life out there, and their future too.

Stephen fell asleep right after his third beer. I don't know what he did on the moon, but I do know that it's a rough place to play at. Thank you, he is not dead.

Chris woke up and went to the bathroom, and then she came into the room and saw Stephen there on the couch passed out.

I was standing near the window looking out to Mars when Criz came over and asked, "Why is he here?"

I told her that he had a bad day on the moon. Cruz laughed and said, "Oh that is a tough place and he survived, wow!"

We both stood there and embraced one another in a hug; we both kissed and smiled at each other. "This is all for the future of the children and our new son Matty" I stated.

Cruz said, "Yeah! For Matty May, our son and the others, we are doing the right thing." We then went to the bedroom to have a long nap and I said softly, "I'm home with love, Criz, good night and good day."

Chapter 13

THE PLANET COUNCIL

It was 10 am when my brother came into our room, he was screaming something like, "We are late, I forgot to tell you, but the council is having an important meeting about Earth and Sohr the bomb. Get up bro we are late."

I jumped off the bed and screamed back, "Why didn't you tell me sooner?"

Stephen said, "I was too drunk to tell you."

I said to him, "What the hell is wrong with you? How could you forget something like this?" I grabbed my pants and shirt and I put them on really fast, I was so happy that I didn't wake Cruz. I got my jump boots on and bounced to the counsel with Stephen by my side.

We just got to the council thirty minutes late. My dad was there; he was talking to Mr. Burt as Stephen walked over to talk to him. Stephen apologized to dad for forgetting to tell me about this meeting.

My dad said, "That's ok my son, just sit back and listen to it all."

I said to my dad, "I'm sorry too" as I sat next to my brother and Mr. Burt.

My dad made a speech: "To all good men, we need to send the bomb Sohr to earth. It's near the two hundred year mark to save life, and this needs to happen soon my followers." Everyone sat up and listened intently.

"In the next two hundred years, the atmosphere will change so that humans can live life on Earth."

They all stood up and cheered with excitement. My dad then sat down and Mr. Burt said, "This is all we needed to say, let the counsel decide when we can start."

Stephen sat up and said to our dad, "When?"

He replied, "As soon as possible, tell John now."

Stephen got up and bounced to John's office to ask him when it would all be ready, then bounced back and sat down in his chair. He was only gone for five minutes. Stephen announced, "I just told John that we need to do this soon and he said it's ready to go."

My dad stood up to the council and asked, "Can I approach this matter?"

The councilmen said, "Yes you may approach."

Graham, my dad, said, "It's ready my lords, we can get this bomb off as soon as possible, like within nine months, or when we are all ready to my lords."

"Great!" they said, "Let it be done" We all applauded and slowly left the room into the foyer, and spoke successfully. No one understood what we were saying.

John was standing there with a big smile and said, "To all good brothers, my experiment is ready."

The people all cried out in joy. John had a tear in his eye. I had never seen this before. John said out loud, "It will work my friends." There was another cry out, this time it lasted longer than the first. What a day to remember. At last, it's ready.

I yelled out loud, "The buzz bomb!" All of a sudden I realized that I had just given Sohr a nickname. The people started to yell out "Buzz bomb!" and it finally got its nickname, thanks to me. John turned around and said to the people, "We have a nickname; we will call it the buzz bomb." So there it is the "buzz bomb".

The people cried out loud "The buzz bomb! Yes! The buzz bomb! Yes!" The people slowly piled out of the building and bounced a few at a time. While John and I were standing there alone, John said, "Well done Simon, I am proud of what you have done my friend."

I replied "Thanks old guy."

We both walked away, and I went for a long walk. We would always look back on this day, as it would go down in history for Mars and its people. Long live the people of Mars.

Chapter 14

THE CELEBRATION

Everything went well at the council meeting, and all of Mars was partying outside. I was there with my family and John was very happy that there was hope for our future. My dad was over in the corner talking to Criz and Stephen with a drink in his hand. Bailey was there too; she left rehab early, just to be with my brother and our family. I had a feeling she liked my brother.

My dad has his hand on my wife Criz's belly and he said to her, "I am glad I have a grandson, but I still feel sad about Mickey.

Cruz said, "Thank you for caring."

I just stood back and watched my family and friends celebrate.

I asked Stephen to go for a walk and he said, "What about a drink?"

I said, "Ok, a drink sounds good." So we went outside and bounced to the nearest bar called, "Home Town Bully's." We walked in and the place was packed. I could see lots of my friends there like Matthias, Markus, and Mike. I greeted them all and they said "Nice work Simon."

I went straight to the bar and asked for a bottle of whiskey. The bartender said, "You back again?"

I said, "Yes" as I handed my brother a glass. "We're here to get drunk and celebrate the new way on Earth."

Stephen went over to these good looking girls and started to talk while I stayed over at the bar and got pushed around some. "There are too many people in your bar," I said to the bartender.

He just said, "This is the biggest party ever here in my bar."

Stephen was over talking with those girls. I need to watch out for him, as he is my brother. I went over to him and one of the girls said: "You're Simon!"

"Yes, that's me," I said with a huge smile on my face "Why?" I asked.

Then one of the girls said, "You're that guy that's found a spot for that buzz bomb."

I said, "Yes, my family and I are involved with the whole thing."

Then another girl said, "I know John."

Stephen said, "You're his niece Emily."

We all laughed as Emily said, "Yes." I felt like a dummy standing there, because she was a good-looking girl. She would be all right for my brother as a girlfriend but he has bailey.

Markus, Matthias, and Mike were starting a fight with some moons, as I walked over there. I had to say something about that. I said "Markus you will lose, they're moons, and they will try to kill you.

Markus said, "I am all drunk up, it won't hurt."

Matthias said, "Back off! This fight is ours."

Mike took the first hit from a moon and was knocked out cold. The fight lasted only two seconds, Markus got punched out and so did Matthias. Everyone in the bar stopped for a few seconds, and then started up again when Stephen said, "Let's party!" really loud.

We all bashed and smashed the place all night. It was a celebration that will last for centuries, in all of our hearts, souls, and minds.

Chapter 15

A PLACE I CALL HOME

It was 4 am, and I snuck in after a long night of drinking. Cruz does not mind when I do this, as long as it doesn't happen all the time. I agree with her wishes; I am a family guy and we know that well.

I went over to the fridge and got out a cold beer, my last one. I went and sat down on the couch and drank some, not too much though, as I'd had a lot to drink last night.

Cruz heard me come through the door and called, "Is that you Simon?"

I said, "Yes my love, it's me, Simple Simon."

She asked, "Are you drunk?"

I answered honestly, "Yes, just a little." as I gasped for some air.

Cruz asked, "Are you done?"

I said, "Yep, all done my love."

Then she said, "Good! Come to bed."

I said, "Just finishing up."

Cruz said, "Ok."

I drank the whole bottle down with one gulp. I then got up and headed to the washroom, as I said, "I'll be there in a minute Cruz."

I finally got to lie down. It had been a long day for me. Cruz started to laugh under the covers, as I took my socks and shirt off.

I whispered to Criz as she giggled, "I need rest."

She said, "No you don't."

I asked, "What do you have in mind?" Criz giggled again and we both went under the covers and made love.

The next morning I was as sick as a dog. I left Chris in the bedroom to sleep. Someone was knocking on the door and I opened it. Oh my God, it was my brother and Bailey, both still drunk.

Bailey asked, "Can we come in?"

I said, "No," but my brother pushed his way in and Bailey followed. They both went for the couch.

I said to them, "I love you guys, but I am sick as a dog; can we do this some other time? They both laughed, "No we are in now; can we chill?"

I sighed, "I guess so, but please don't wake Criz; she is in the other room."

Bailey said, "No problem, Simon," as her head hit the pillow behind her.

"Simon!" Stephen said.

"What?" I replied.

Stephen said, "Can I crash here too?"

I said, "Yes you can, just don't wake Criz, she will freak."

Stephen's head hit the pillow as he said, "I am done."

I then got up and went back to the bedroom. Cruz was still sleeping. I left her like that, looking like a cat.

"How cute," I said softly as I kissed her on the forehead. Then I went back to the kitchen for a glass of water.

Cruz started to move and moan a little as I walked back into the bedroom. "Cruz," I said, "Criz are you ok?" I asked.

She said, "Yeah I am ok."

I asked her, "Would you like a drink?" Cruz said softly, "Yeah, please."

So I went and got her a glass of cold water. "Thanks," she said when I returned.

Just then there was another knock on the door, and my dad Graham walked in. "Boys, the bomb will be ready in nine months," he said, as Stephen woke up in shock to my dad standing there at the door.

I said, "Cool! Can't wait for dad." Bailey was still passed out.

Cruz heard from the bedroom, and said out loud, "Yes it's true, Graham told me yesterday, after you guys left."

I said, "Wow, this is all good, it will happen sooner than I thought."

Everyone was up after that announcement and talking about the bomb. We are all in the den, Criz, Stephen, Bailey and my dad.

I stood outside looking at our red atmosphere and said, "Matty May, and this is for you and Mickey."

Cruz came out and she gave me a big hug and said, "I miss my dead son, I am glad we will have a new life with a new child."

I said, "Yeah me too. I can't wait to see the buzz bomb off. Next week or nine months later. It will happen soon".

Chapter 16

THE VISIT

Well, it had been a while since we had seen my in-laws, and I wasn't too happy to go, but Cruz had been all packed up to go for a day or two. I thought about staying; I was not looking forward to going.

We were ready to go on short notice and we had our bounce boots on. We made it to the station and were waiting in line for the jump. All of a sudden we are off, going through the wormhole. Within fifteen minutes we were at the end.

When we got to the end, we bounced right to their front door. Chris rang the bell, and their butler answered the door. I said to Cruz, "Let's not stay too long."

She whispered, "Ok we won't."

Just then the butler stated in a deep voice, "Your mom and dad, Jag and Mary Day are on vacation."

Cruz asked, "Where?"

The butler replied, "Some distant planet, far off."

I took a gasp of air and politely asked, "Cool, can we stay the night? "As Criz and I walked right in.

It was a big house with five bedrooms, five bathrooms, and a huge kitchen and basement. They also had a huge TV, which I planned on watching later.

Cruz went up to see her old room; she is the only child and has lots to do in the house. The butler asked if I would like a drink and I replied, "Yes please, a beer." and he went off to get it. I spoke out loud, "This is a lot better without the Days."

There was no one around to hear me say that to myself, "Great stuff!" as I sat on the couch. I was alone, so I put the TV on, put my feet up on the table and relaxed with my beer.

I finished about six beers, and I fell asleep with the TV on. I woke up the next morning about nine o'clock. Cruz was still upstairs asleep, in her old bed. I walked up the stairs and walked into her room, "What a cutie." I said.

Cruz just moaned a little and rolled over with a little smile on her face. I went back downstairs and wrote a letter to her that read,

I am leaving, I need to get back to earth, and I have a few things to do for myself. Love you, and the baby Matty, take care, my love.

Chapter 17

THE FIGHT

It was in the afternoon, and I was working in the kitchen making Criz something to eat when the doorbell went rang. I walked over to the door and opened it, it was my brother Stephen holding tickets in his hand, and he said. "Simon, I have tickets to a fight, would you like to go?"

I said, "Yes when?"

Stephen said, "Now, let's go, hurry up, killer Keller from the rehab on Venus, is the title wait fighter, I've got one thousand dollars on him to win with one round."

So I went into the bedroom and told Criz that I was going out for a while with my brother Stephen.

She said. "Ok, have fun." I started to get dressed to go.

Stephen waited, as I got dressed. He was excited to go, so was I. This was going to rock!

We both bounced to the forum and walked in, and we grabbed our seats, just near ringside. What a view, there were a bunch of people partying behind us; this felt good as I just sat there with excitement.

Stephen was drinking a beer and he said, "This is going to be the best fight ever."

I agreed, "Yes so far so good." I laughed and I took a bite from my hotdog.

Stephen said, "Five minutes to go Simon, five minutes, yeah!"

The lights went down and the announcer came out onto the ring, and said, "We are here today to witness the allotment fight of the century. In the left corner we have Killer Keller, two hundred pounds and takes his name well; in the right corner, we have Vulgar, two hundred and five pounds and standing champ. Let the fight begin. They hit their hands together, and they both started to dance, for the fight.

Killer came out dancing, and gave vulgar one good shot to the head; he knocked him out, just like that! Stephen stood up, and gave out a party cry and said, "Wow! I am rich! I told you, brother, things pay off." The whole stadium rocked with screams with yells; it was so loud, I had to cover my ears.

The announcer came out, and took Killer's hand and waved it in the air, and said, "We have a champ."

The crowd went nuts, Stephen and I jumped up with excitement with our arms in the air screaming, "Killer," over and over again; the crowd did not stop screaming, "Killer Keller!"

Stephen just got up and left, so he could cash in his winnings, as I wandered around the stadium. For a while, I was in shock, as I thought, *what a fight, short and sweet, yelling out with all the other people, "Killer Keller."* I said this once in a while, to join in with the crowd. What a party!

Chapter 18

THE BABY SHOWER

It was five in the afternoon, and I was just leaving John's to be home; I bounced right to my front door and walked in.

Cruz was there with a bunch of her friends, it looked like a party, so I asked, "What do we have here?"

Cruz said, "A baby shower dear."

I said, "Sorry to crash it, but I'm hungry," as I took some food off a plate, and shoved it into my mouth. I said, "Thanks bailey."

Bailey looked up and smiled, and asked, "What for?"

Donna a friend of Criz was looking at me, and she smiled at me, while Jasmine giggled.

I said, "Hi Mary Day love you too." Mary just looked at me with her eyes and said, "Love you too silly boy."

Chris asked, "How is Stephen?"

I replied, "He's ok. I left him at John's with the jug bugs."

Cruz said, "Cool just leave him there."

I replied, "Ok," as I went to the fridge for a drink, and walked outside on the patio and looked up at the red sky in fright that this planet might be our last.

I could barely hear the girls outside, as I sat down on a chair and dazed off.

Cruz was having fun and she loved her friends just a girl thing.

When I woke up, the girls were starting to leave. This made me think it was myself. So I got up and went to the door and asked, "Is everything ok, it's not me why you girls are leaving, is it?" All the girls said in a different time "No way, not you, no way Simon" as they left through the front door with giggles.

Cruz and I were finally alone, so I asked, "What is all this?"

Cruz said, "Gifts from our friends."

I said, "Wow look at this," as I held up a tiny T-shirt.

As Chris tried to grab it out of my hands I said, "What? Isn't it mine?"

Cruz laughed. She went to the fridge for some water and opened it.

I sat down on the couch and Cruz said, "That was a nice party; I enjoyed it all with my friends."

I said "Nice baby. Love you." As I closed my eyes, Criz went to the bedroom to get into something nicer.

Two hours later I woke up, with Criz lying with her head on my lap. She said, "Hi Simon you're with me now."

I asked, "Are we safe?"

Cruz said, "You're at home."

I asked, "Are we ok?"

Cruz said, "Yes my dear, we are ok."

"How is the baby?"

Cruz said, "Any day now sweetie." As I felt Cruz's belly with my hand and gave her a long kiss with love. We both got up and went into the bedroom to get some rest. We fell into a deep sleep to rest up for the birth of Matty May.

Chapter 19

THE BIRTH

A few months later it was five in the morning and I was at the fridge, and Criz was in the bedroom. All of a sudden she cried out, "My water broke!"

I came running in and asked, "Are you ready?"

She said, "Yes let's go to the hospital now."

I got her bag and I made a call to her doctor and said, "The Mays is ready." Within minutes someone had knocked on the door and I opened it, it was the doctor and a few nurses. They had bounced together and got there really fast. They got Criz ready for birth, and the doctor asked me if I could stay. I said, "Ok."

With all the technology the birth happened fast; it took ten minutes to deliver the new baby. They gave Criz some meds and used probes to help the birthing process go fast. I could not believe that my boy was finally born. I walked over and saw him all bloody and stuff, and I said, "Hi! This is your mom."

Cruz started to cry and asked, "Is he healthy?"

The doc said, "Very healthy boy for his mother and family to love."

I said, "Yes."

About half an hour went by and people started to show up to have a look at Matty May, our newborn. My dad was there and so were Stephen and Bailey. John took his time and finally showed up.

Our place was full of people from the conference, most of them I only know from seeing their faces. There was one person I recognized and her name was Jasmin, but I didn't know her last name. Jasmin came over to me and I said, "Hi Jasmin."

She said, "Congratulations Simon on your newborn, and it's near the launch. You must be a very happy father."

I said, "Yes, I am," as she kissed me on the cheek and hugged me. Jasmin giggled a little bit, then walked away to get a drink from the bar. Jasmin was one of Cruz's old friends from high school, on Pluto.

Cruz was still in the bedroom with the nurse, exhausted and resting with Matty in her arms. I gently pulled the door closed so I could entertain our guests.

This whole thing reminded me of the birth of Mickey and the death of Val Snow, my mother. It was a good thing that I kept it to myself, 'because if Criz knew what I was thinking, she might slip into a depressed stage too soon after the birth.

Cruz's family was there too, but didn't say much to me, as they wanted her to give birth on Pluto. What a shame it is, as Pluto is her home planet. Her dad Jag, made me a gesture with his glass and looked away fast while carrying on in conversation with his wife.

All of a sudden my dad Graham lifted his glass for a cheer and said, "To Matty, my new grandson." Everyone replied "Here, here, to Matty!" and drank some Champagne from tall glasses.

My brother said to all of the people, "Life is life. So enjoy it the best you can my brother and friends." That was a cool thing to say, I enjoyed that speech a lot.

So I said, "Thanks Stephen, I agree to that," while holding my drink in the air. I took a big long drink of Champagne.

Bailey walked over and hugged me and whispered in my ear; "I wish I could have a baby with Stephen." Then she walked back across the room and stood there beside my brother.

I finally decided to relax in the room with Criz and my new son, while the party went on in the other room. I asked Chris how she was doing, and she said, "I am ok, how is Matty?"

I replied, "He is beautiful my love," as I sat in a chair beside the bed. Cruz was lying down and I held her hand as I fell asleep by her side, with Matty in her arms. The party was over for me. Just two more days till the launch.

Chapter 20

THE LAUNCH

It took nine months, but the launch was finally ready. All of my family was out on our patio looking up to the sky, just looking at the red sky and the damage we had done to Mars. We were only human, as some people would say, but we learned from our mistakes.

Mars is now a sick planet just like the others; we were foolish for destroying it the way we had. We were greedy and stubborn, and when it comes to the truth; we lie and cheat like there is no tomorrow, trying to get ahead in any way. Mars will always be in the sky, to remind us how we should live.

The bomb was ready, just a few seconds more to go. All of a sudden there was a loud noise, sounding like a bunch of rockets, and a bolt of light in the sky. It was off, heading to earth.

All the people cheered like they were at a football game when a goal had just been scored. In only five hours it would hit earth and we would see the Milky Way light up.

Five hours later, as predicted, the sky lit up, the buzz bomb hit earth, and the sky lit up for about twenty seconds in ten short blasts. Life had just begun and all the people were out screaming in the streets, thanks to John and all of our family. Life was on the move and would last for some time.

That day on earth, we asked the elders not to speak about what happened to Mars and the other planets. The secret would be kept hidden and we would place these people far away from each other so that their new life would grow slowly, and away from the truth. Life on Earth had to last a long time; it's the last planet left, or is it?

My baby's boy's boys will have kids, which may be able to live on earth with the elders. They were giving the atmosphere two hundred years to settle before anyone could step on Earth.

The bomb would change the atmosphere, so nothing will live; it will kill off all of life. The bombs will also send the earth spinning in an odd circle, so it will get cold and hot just like winter and summer. Earth will gradually heal itself, and after trillions of years, humans will not be able to live on Earth. For the next two hundred years, we have time to plan, like what will we plant and take to earth?

As long as we are still alive and well, we will always find something new that will destroy something old. We are only human, and we are always in a race to make something new. We fix the broken and replace the old, for new things that will hurt our home. What else can we do but live and learn new ways and forget what we did to Mars. Not a chance! This time let's slow down and listen to the earth. It cries out, "Mars! I miss you, I am next what shall I do?" as life goes on, because after all, we are only human and we kill, "THE NEW EARTH"

"A CHANCE TO SOHR"....... "MARSANIA"

BY GARWICK STEVENS

ISBN: 1500311278
ISBN 13: XXXXX
Library of Congress Control Number: XXXXX (If
applicable)
LCCN Imprint Name: City and State (If applicable)

www.ingramcontent.com/pod-product-compliance
Lightning Source LLC
Chambersburg PA
CBHW071630170526
45166CB00003B/1263